WHAT IS SQUARE?

In memory of *Little Grandma*,
Ruth Kelly Thompson, who was anything but square.
—R. K. D.

Per Zio Michele
In memory of my uncle, Michael Orlando,
whose gentle smile and magic tricks
filled my childhood with laughter and awe.
—M. F.

Thanks to Joy Nagy for painting the backgrounds, to Kate Carlyle of Mustard Seed and Moonshine for the square teacup, and to Cortland Kerecz and Gabriel Peter Mollica for showing us the patches on their knees.

ISBN 0-439-15945-8

12 11 10 9 8 7 6 5 4 3 8 9 10/0

Printed in the U.S.A. 23

First Scholastic printing, March 2000

Design by Tom Starace

WHAT IS SQUARE?

by REBECCA KAI DOTLICH
photographs by MARIA FERRARI

SCHOLASTIC INC.
New York Toronto London Auckland Sydney
Mexico City New Delhi Hong Kong Buenos Aires

A biscuit, a basket,

a patch on
your knee,

a white sugar cube for a square cup of tea.

A colorful
quilt,

a cold cube of ice,

caramels and crackers,

a dishcloth, and dice!

A window, a waffle,

a tin baking pan.

Can you find a square button?
Of course you can!

One checkerboard,

two alphabet blocks,

a pillow,

a puzzle,

a buckle, a box.

A mirror, a map,

a brownie,

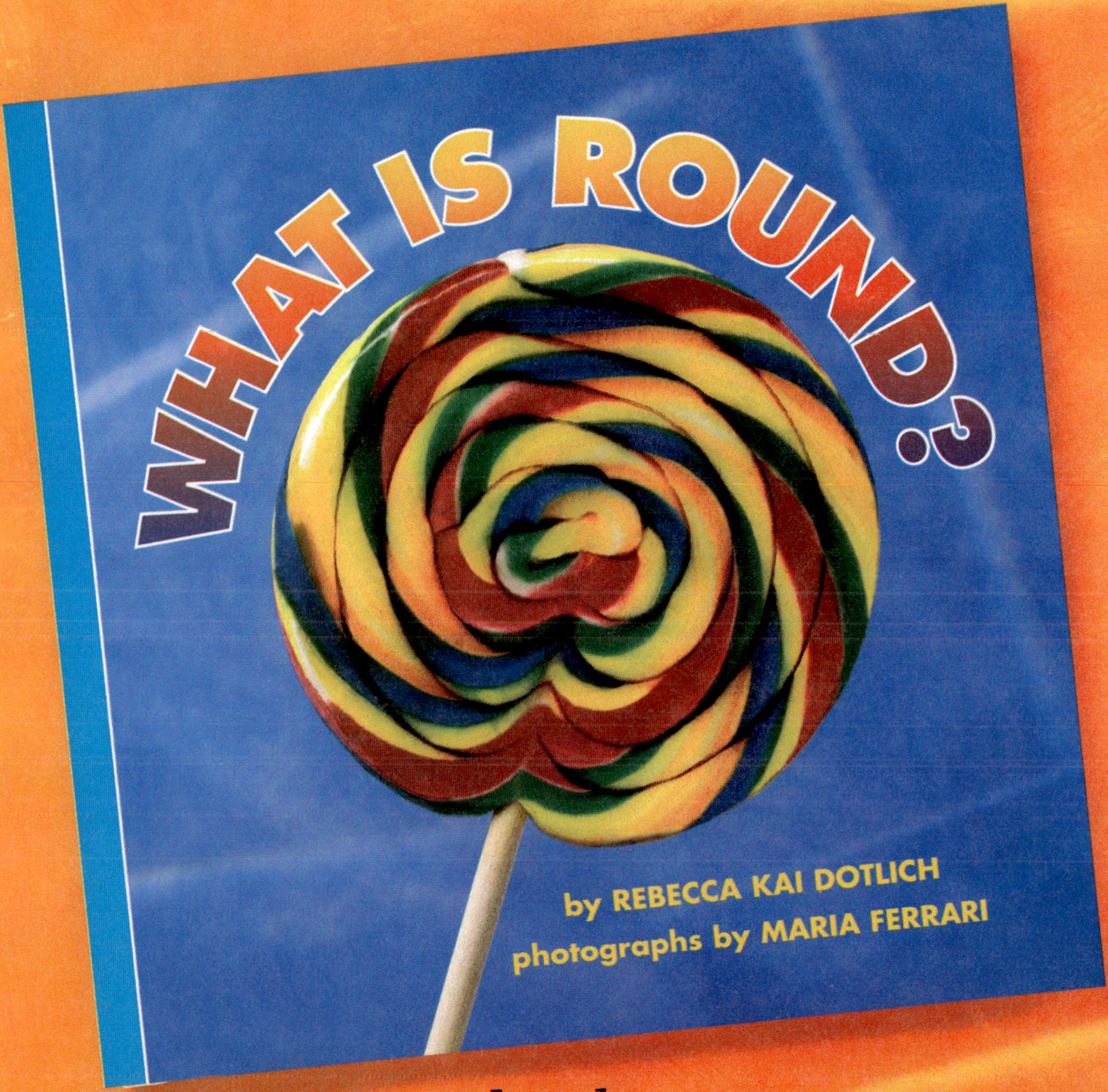
WHAT IS ROUND?
by REBECCA KAI DOTLICH
photographs by MARIA FERRARI

a book. . . .

What else is SQUARE?

Your turn to look!